Full STEAM Ahead!

Math Matters

Data Is Information

Field Trip

Farm	😄 😄 😄 😄
Museum	😄 😄 😄 😄 😄
Science center	😄 😄 😄 😄 😄 😄

😄 = 1

😄 = 2

Adrianna Morganelli

CRABTREE
PUBLISHING COMPANY
WWW.CRABTREEBOOKS.COM

Title-Specific Learning Objectives:

Readers will:

- Define data and give examples of different types of data.
- Read and understand tally charts, pictographs, and bar graphs.
- Use the charts and graphs in the book to describe how data can be displayed.

High-frequency words (grade one) a, and, by, he, is, make, the, they, to	Academic vocabulary bar graph, pictograph, tally chart

Before, During, and After Reading Prompts:

Activate Prior Knowledge and Make Predictions:

Ask children about a relatable topic, such as eye color, and record the data from their responses. Have children raise their hands if they have blue, brown, green, or hazel eyes. On a whiteboard or chalkboard, write down the number of children who have eyes of each color. Do not erase the numbers as they will be used for an activity after reading the book.

Tell children that these numbers are called data. Then, have children read the title of the book and look at the cover image. Ask children if they can give any other examples of data. Discuss what children think this book might be about.

During Reading:

After reading pages 13 to 15, ask children to look at the different ways the data was displayed. Ask them:

- How did Mateo and Shreya display the data?
- How are the pictographs on pages 14 and 15 similar? How are they different?

After Reading:

Direct children's attention back to the data that was collected before reading. Have them work in groups to display the data using one of the methods from the book: tally chart, pictograph, or bar graph. Have each group share their chart or graph after completion.

Author: Adrianna Morganelli

Series Development: Reagan Miller

Editor: Janine Deschenes

Proofreader: Melissa Boyce

STEAM Notes for Educators: Janine Deschenes

Guided Reading Leveling: Publishing Solutions Group

Cover, Interior Design, and Prepress: Samara Parent

Photo research: Janine Deschenes and Samara Parent

Production coordinator: Katherine Berti

Photographs:
iStock: omgimages: p. 6 (t); Vladimirs: p. 12; Nenov: p. 13 (r)

All other photographs by Shutterstock

Library and Archives Canada Cataloguing in Publication

Title: Data is information / Adrianna Morganelli.
Names: Morganelli, Adrianna, 1979- author.
Description: Series statement: Full STEAM ahead! | Includes index.
Identifiers: Canadiana (print) 20190231815 |
  Canadiana (ebook) 20190231823 |
  ISBN 9780778772378 (hardcover) |
  ISBN 9780778772712 (softcover) |
  ISBN 9781427124647 (HTML)
Subjects: LCSH: Statistics—Juvenile literature. |
  LCSH: Statistics—Graphic methods—Juvenile literature.
Classification: LCC HA29.6 .M67 2020 | DDC j001.4/22—dc23

Library of Congress Cataloging-in-Publication Data

Names: Morganelli, Adrianna, 1979- author.
Title: Data is information / Adrianna Morganelli.
Description: New York, New York : Crabtree Publishing Company, 2020.
  Series: Full STEAM ahead! | Includes index.
Identifiers: LCCN 2019052996 (print) | LCCN 2019052997 (ebook) |
  ISBN 9780778772378 (hardcover) |
  ISBN 9780778772712 (paperback) |
  ISBN 9781427124647 (ebook)
Subjects: LCSH: Mathematical statistics--Juvenile literature. |
  Information visualization--Juvenile literature.
Classification: LCC QA276.13 .M67 2020 (print) |
  LCC QA276.13 (ebook) | DDC 372.7/9--dc23
LC record available at https://lccn.loc.gov/2019052996
LC ebook record available at https://lccn.loc.gov/2019052997

Printed in the U.S.A./032020/CG20200127

Table of Contents

Crabtree Publishing Company
www.crabtreebooks.com   1-800-387-7650
Copyright © **2020 CRABTREE PUBLISHING COMPANY**. All rights reserved. No part of this publication may be reproduced, stored in a retrieval system or be transmitted in any form or by any means, electronic, mechanical, photocopying, recording, or otherwise, without the prior written permission of Crabtree Publishing Company. In Canada: We acknowledge the financial support of the Government of Canada through the Book Publishing Industry Development Program (BPIDP) for our publishing activities.

Published in Canada
Crabtree Publishing
616 Welland Ave.
St. Catharines, Ontario
L2M 5V6

Published in the United States
Crabtree Publishing
PMB 59051
350 Fifth Avenue, 59th Floor
New York, New York 10118

Published in the United Kingdom
Crabtree Publishing
Maritime House
Basin Road North, Hove
BN41 1WR

Published in Australia
Crabtree Publishing
Unit 3 – 5 Currumbin Court
Capalaba
QLD 4157

A Special Gift

Today is Mateo's birthday! His parents give him a special gift. It is a stuffed lion to add to his stuffed animal **collection**!

Do you have a collection? What do you collect?

4

Mateo adds the lion to his collection. He looks at his animals. There are lions, tigers, and hippos. He sees monkeys and cheetahs too! "How many animals of each type do I have?" he wonders.

First, Mateo needs to **sort** his animal collection. He makes groups of each animal type.

Gathering Data

To answer his question, Mateo **gathers** data. Data is a collection of facts about something. There are different kinds of data.

Numbers are a type of data.

Measurements are a type of data.

Observations and descriptions are data too. These students write down what they observe in nature.

The number of animals in Mateo's collection is data. He makes a tally chart to **display** the data. A tally is a mark on a piece of paper. You can find a total number by counting the tally marks.

Animal Collection

Lion	卌 ||
Tiger	||||
Hippo	||
Cheetah		
Bear	卌

The title tells a reader what data is being displayed.

The fifth tally mark is drawn across the first four marks to make a group of five. This makes it easier to count the tally marks.

Mateo has seven lions and four tigers. He has two hippos and one cheetah. How many bears does he have?

7

Popular Animals

At school, Mateo shows his tally chart to his friends. "I like animals too," says Shreya. "Dogs are my favorite." Mateo has an idea. "Let's collect data about the favorite animals of our friends!" he says.

Mateo and Shreya ask each person which animal is their favorite. They make a tally mark for each answer.

Favorite Animals

Dog	IIII II
Lizard	III
Horse	III
Cat	IIII IIII
Lion	II
Koala	I

Mateo and Shreya find that cats are the most popular animal. The animal with the most tally marks is the most popular.

Which animal is the least popular, or has the least tally marks? Which animals are equally popular, or have the same number of tally marks?

Sorting at Recess

Mateo, Shreya, and their friends love to play games at recess. Some children play basketball. Others kick soccer balls into nets.

"Let's collect data about all the balls in the playground!" says Shreya.

Mateo and Shreya sort the balls by each type. They count each group and use a tally chart to display the data. Which type of ball is the most popular?

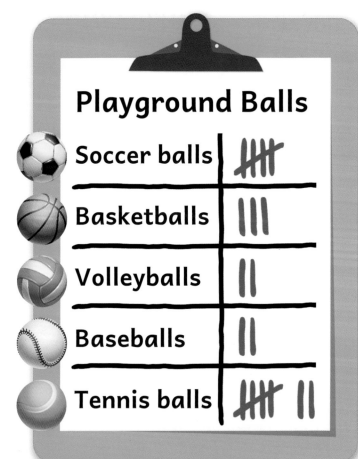

Playground Balls

Soccer balls	卌
Basketballs	|||
Volleyballs	||
Baseballs	||
Tennis balls	卌 ||

"I wonder if there is a different way to sort the balls," says Mateo. Shreya has an idea! "We can sort the balls by their size."

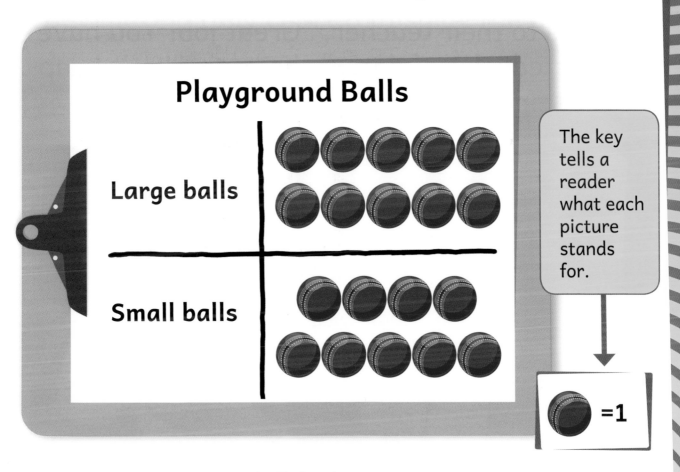

Playground Balls

Large balls

Small balls

The key tells a reader what each picture stands for.

= 1

Mateo and Shreya sort the balls by their size.
This time, they make a **pictograph**. Each picture stands for one sports ball. Which size of ball is the most popular?

Different Pictographs

"Try sorting the blocks in a new way," says the teacher. Mateo decides to sort the blocks by their shape. He makes a pictograph to display the data.

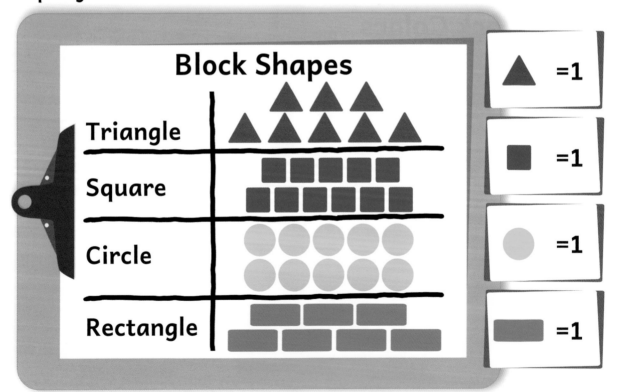

How many blocks of each shape are there? How do you know?

Shreya decides to sort the blocks by their size. She makes a pictograph too. It is different from Mateo's pictograph. Each of Shreya's pictures stands for two blocks.

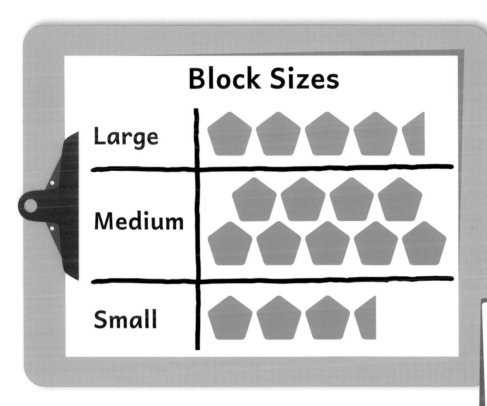

Block Sizes

Large	
Medium	
Small	

Shreya's key says that one picture stands for two blocks. It also says that half of a picture stands for one block.

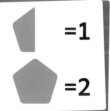

= 1

= 2

Can you use the chart to count the number of large blocks? How about the number of medium blocks and small blocks?

All Kinds of Data

At lunch, Mateo shares birthday cupcakes with his classmates. He thinks of more data to show. "Let's find out which cupcake flavor is the most popular," he tells Shreya.

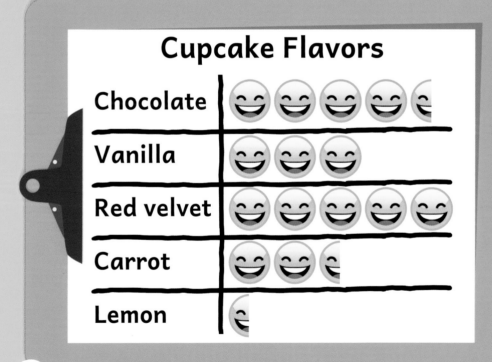

Cupcake Flavors

Chocolate	😄😄😄😄😉
Vanilla	😄😄😄
Red velvet	😄😄😄😄😄
Carrot	😄😄😉
Lemon	😉

😉 =1

😄 =2

Which cupcake flavor is the most popular? Which is the least popular?

"What other data can we collect?" asks Shreya. The friends decide to ask their classmates about their favorite subjects.

Favorite Subject

Language	卌
Art	卌 卌 Ⅱ
Math	卌 Ⅱ
Science	卌 Ⅱ

Which subject is the most popular? Which subjects are equally popular? How could Mateo and Shreya display this data in a pictograph?

Making Bar Graphs

After lunch, the children learn about a new way to display data. "A **bar graph** is a good way to show data," explains the teacher.

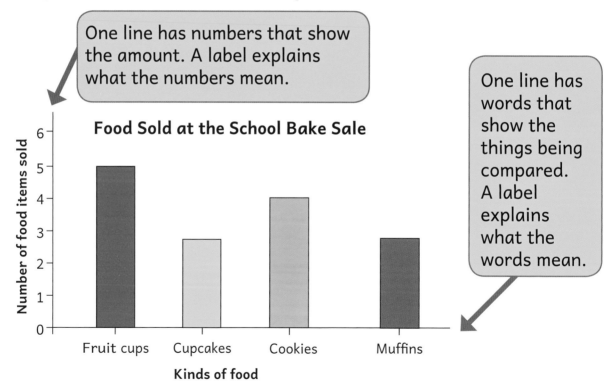

One line has numbers that show the amount. A label explains what the numbers mean.

One line has words that show the things being compared. A label explains what the words mean.

Food Sold at the School Bake Sale

Number of food items sold

Fruit cups   Cupcakes   Cookies   Muffins

Kinds of food

A bar graph uses rectangular bars to show data. This is a **vertical** bar graph. The highest bar shows the most popular food. The lowest bar shows the least popular food.

Mateo and Shreya decide to practice. They make a bar graph with the data they collected about school subjects. "The graph makes the data easy to understand!" says Shreya.

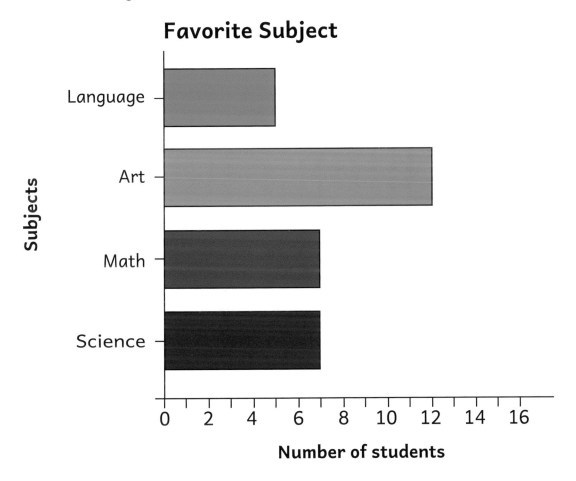

Favorite Subject

This is a **horizontal** bar graph. The longest bar shows the most popular subject. The shortest bar shows the least popular subject.

A Lot to Learn

Mateo, Shreya, and their classmates make more graphs to display data. Their teacher tapes the graphs in the school hallway to share the data. The students are proud of what they have learned!

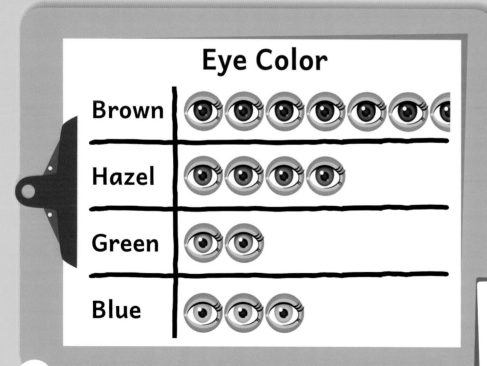

Eye Color

This pictograph displays data about the eye colors of the students in the class. How many students have each eye color?

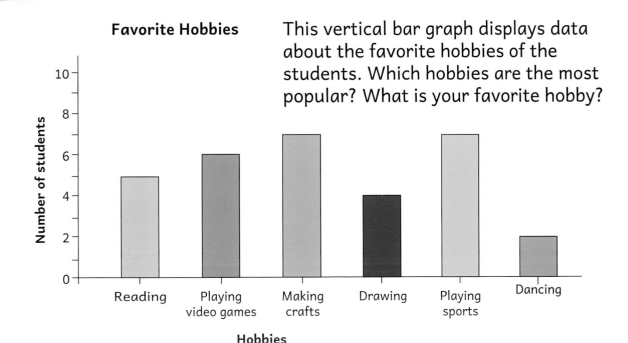

Favorite Hobbies

This vertical bar graph displays data about the favorite hobbies of the students. Which hobbies are the most popular? What is your favorite hobby?

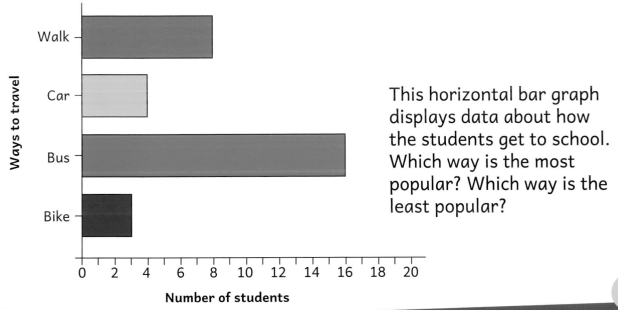

Ways to Get to School

This horizontal bar graph displays data about how the students get to school. Which way is the most popular? Which way is the least popular?

Words to Know

bar graph [bahr graf] noun A graph that uses horizontal or vertical lines to display data

collection [kuh-LEK-shuhn] noun A group of items that is kept together in one place

display [dih-SPLEY] verb To show or make possible to see

gathers [GATH-ers] verb Brings more than one object together into one group

horizontal [hawr-uh-ZON-tl] adjective Straight across

pictograph [PIK-tuh-graf] noun A graph that uses pictures or symbols to represent data

sort [sawrt] verb To arrange a group of objects in a certain way, usually by similar characteristics

vertical [VUR-ti-kuhl] adjective Straight up and down

A noun is a person, place, or thing.

A verb is an action word that tells you what someone or something does.

An adjective is a word that tells you what something is like.

Index

About the Author

Adrianna Morganelli is an editor and writer who has worked with Crabtree Publishing on countless book titles. She is currently working on a children's novel.

To explore and learn more, enter the code at the Crabtree Plus website below.

www.crabtreeplus.com/fullsteamahead

Your code is:
fsa20

STEAM Notes for Educators

Full STEAM Ahead is a literacy series that helps readers build vocabulary, fluency, and comprehension while learning about big ideas in STEAM subjects. *Data Is Information* uses pictures, charts, and graphs to help readers gather and understand information about data and how it is displayed. The STEAM activity below helps readers extend the ideas in the book to build their skills in math and science.

Learning from Data

Children will be able to:
- Collect and display data about classroom waste.
- Use the data to help create a classroom waste-reduction plan.

Materials
- A full or semi-full garbage can, depending on its size, a recyclable floor covering, and rubber gloves
- Chart paper and markers
- Waste Data Worksheet

Guiding Prompts

After reading *Data Is Information*, ask children:
- What are some types of data?
- How do charts and graphs help people understand data?

Activity Prompts

Explain to children that collecting data can help us solve problems because it lets us learn about the problem. Explain that today, we will collect data that will help us solve a big problem: people create too much waste.

Tell children that first we need to collect data about the waste we create in our classroom. Empty the garbage can onto the floor covering. Have children put on rubber gloves and, together, sort the waste into different groups. Then, on a piece of chart paper, create a tally chart with the number of waste items in each group. (Groups could include such categories as food, paper, cardboard, plastic wrappers, straws, plastic bottles, plastic containers, glass bottles, etc.) Clean up the waste safely.

Have children work in small groups to complete the Waste Data Worksheet. Instruct them to use the data from the tally chart to make a new tally chart and bar graph. This time, the data is sorted in a new way: by how the waste could be disposed. Review definitions of compost, reuse, and recycle.

Have each group share their data with a new group. As a class, discuss how you can work together to reduce the amount of waste you create. Make a class plan to solve the problem.

Extensions
- Children can design posters that communicate the class waste-reduction plan. The posters should include data from the activity.

To view and download the worksheets, visit **www.crabtreebooks.com/resources/printables** or **www.crabtreeplus.com/fullsteamahead** and enter the code **fsa20**.